THE BIOMEDICAL RESEARCH INSTITUTE: A PERSPECTIVE ON ALTERNATIVE PARADIGMS FOR RESEARCH DEVELOPMENT

Paul E. Neiman, M.D.

Member and Director Emeritus
Division of Basic Science
Member Emeritus
Division of Human Biology
Fred Hutchinson Cancer Research Center

Emeritus Professor of Medicine
University of Washington

Fred Hutchinson Cancer Research Center
Seattle, WA

Fred Hutchinson Cancer Research Center, Arnold Library Cataloging-in-Publication Data

Neiman, Paul E.
The biomedical research institute: a perspective on alternative paradigms for research development / Paul E. Neiman. – Second Edition.
22 p.
Includes bibliographical references.
ISBN 978-0-945278-07-8 (paperback)
1. Biomedical organizations. 2. Medicine—Research. I. Neiman, Paul E. II. Title.
[DNLM: 1. Biomedical Research—organization & administration. 2. Biomedical Research—economics. 3. Models, Organizational. 4. Faculty. 5. Research. W 20.5]
A digital copy of this is also available in the Fred Hutchinson Arnold Library repository at http://authors.fhcrc.org/1101

Cover photo contributed by Ron Reeder, Ph.D.

ACKNOWLEDGEMENTS

The author is grateful to Ann Marie Clark, Head Librarian and Director of the Arnold Library of the Fred Hutchinson Cancer Research Center for essential assistance with publication, and to Ann Reynolds, Staff Scientist, and Miriam Wnuk of the Library staff for excellent editorial assistance.

CONTENTS

Foreword

Like most scientists moving to their first faculty position, I had no concept of how a faculty functions as a social and political body. When I arrived at the Fred Hutch in Seattle in 1985, I was already aware that the Basic Sciences Division had amazing science and strong personalities, but no idea that five years previously there had been a major upheaval, a rebellion even, that had essentially replaced a top-down system, in which senior faculty recruited junior faculty into their programs, with an anarchic, grass-roots system where only the quality of science mattered and individual faculty members were equals.

That revolt precipitated the replacement of the scientific director, the division of the center into separate arms representing public health, clinical research and basic science, and the selection of Paul Neiman, one of the junior lab-based faculty, to be the first head of the new Division of Basic Sciences. I sense that Paul realized that his new role would take him away from his research, but he took it on with gusto, working with other young faculty to create a system that would be as egalitarian as possible.

The idea was that a flat, non-hierarchical structure would minimize rivalries over space or salary and provide every member with an equal voice. The result they hoped for, and achieved, was one of mutual respect and support, and a common goal of maximizing the success of the center as a whole.

The timing was just right. The molecular biology revolution brought new tools for manipulating DNA, opening up many new areas for research. There were so many questions to answer, and hallway conversations were common. There was a real buzz of excitement. By the time that I arrived as the most junior faculty member, the scheme was already working and has been followed, albeit imperfectly, ever since.

More than thirty years later the Basic Sciences Division still adheres to the principles established by Paul and his colleagues in those early days. Equal space and salary, equal voice, help those that need it, promote the science. From each according to his ability, to each according to his need is increasingly hard to apply, but still worth aiming for. Compromises are inevitable, but we try our best. Our underlying philosophy is a big selling point when we recruit new faculty members, because they see the benefit of an apolitical structure and know their opinions matter. Moreover, the junior faculty are motivated to carry the tradition on into the future as the old guard retires.

After he retired, Paul wrote a history of the Basic Sciences Division, including the transition from top-down program structure to the current system. That process led him to reflect on the two systems, and to pen the current Perspective, in which he contrasts the flat, egalitarian model against the more standard hierarchical model.

As the current division director, it's been a privilege to work with a faculty structure that was shaped by Paul, and I'm pleased that he asked me to write a foreword. I believe that Paul's vision can guide other research institutes as they balance program agendas against pure investigator-initiated research. Program structures often win out, driven by the perceived need for synergies and group science, but this text lays out a strong case for independence and equality, as it is practiced today at Fred Hutch.

Jonathan A. Cooper, Ph.D.
Member and Director, Division of Basic Sciences
Fred Hutch, October 2015

PREFACE

In 1972 the Fred Hutchinson Cancer Research Center (FHCRC) in Seattle, Washington opened its doors, and in a few short decades developed from the dreams and ideas of its founders into one of the world's leading biomedical research institutes. This uniquely rapid record of accomplishment was not underwritten by a major philanthropic endowment, nor driven primarily by high profile celebrity leadership. In an attempt to record how the FHCRC (nicknamed "the Hutch") came to be, and how the laboratory-based basic research program of the Hutch grew, Barbara L. Berg and I initiated the FHCRC History Project. This initial effort took the form of two monographs, deposited in a nascent institutional archive, covering these aspects of the Center's development in its first period of operation, 1972 to 1996.

Fifteen years after the period covered by the History Project, I am returning to the task to conduct a less parochial exercise still largely derived from personal observations and my understanding of the development of the FHCRC scientific enterprise. In this case, I wish to generalize from one cornerstone of the success of that adventure: the nurturing of a widely admired and effective scientific culture. Rather than dwell further on history per se, the attempt here is to describe some principles of scientific program development that can be illustrated in particular cases by the success of the Hutch.

I began with a search for literature on scientific program development, and the effects of institutional organizing principles on successful development. Elliot C. Kulakowski and Lynne U. Chronister in their recent 900 + page tome on Research Administration and Management[1] provide ample descriptions of the myriad technical aspects of administering an academic research enterprise. A short article by Louis G Tornatzky and Paul G. Waugman, in this extensive reference work, focuses on the role of senior leadership in promoting faculty research within competing activities of a University (e.g. teaching and community service)[2]. Their article centers on fostering a commitment to research at a full service academic institution (e.g. a college of arts and sciences or a school of medicine) that aspires to an enhanced research portfolio.

It has been recognized for nearly a century that scientific progress in the U.S. is not uniform. Achievement tends to be focused in time and place[3]. Addressed to this fact, and at another pole of this spectrum in the literature, the work of J. Rodgers Hollingsworth and colleagues dwells on over-arching features of the sociology of science and scientific institutions that impact transformational scientific progress[4]. These authors worked backwards from a long list of major Nobel Prize-level discoveries in an attempt to reveal general characteristics of the scientific environment in which major discoveries, arose, and which may make such achievements more or less likely. They discuss such general features as institutional diversity and flexibility, and the effects of commercialization of research, over time in the US and internationally.

The scope of my perspective falls somewhere between the poles of this spectrum of literature. My experience derives from a research institute in which the commitment to scientific research is primary, and is essentially, the whole program of the institution. That said, much of what will be discussed here is also relevant to development in a research-intensive university. At this writing a list of articles in Wikipedia encompasses hundreds of medical and/or biological research institutes, concentrated, but not exclusively, in North America and Western Europe. Therefore biomedical research institutes, like the Hutch, are a substantial component of the scientific enterprise world-wide. As mentioned, experience with their development may be expanded usefully to apply to research-intensive universities, academic departments, and other elements within them, that function essentially as research institutes.

Paul E. Neiman, M.D.

Member and Director Emeritus
Division of Basic Science
Member Emeritus
Division of Human Biology
Fred Hutchinson Cancer Research Center

Emeritus Professor of Medicine
University of Washington

THE BIOMEDICAL RESEARCH INSTITUTE: A PERSPECTIVE ON ALTERNATIVE PARADIGMS FOR RESEARCH DEVELOPMENT

A core objective for administration of a biomedical research institution is the recruitment, career development and material support of talented and productive faculty investigators. The extent to which success is achieved in this effort will reflect the impact of this faculty, individually and collectively, on progress in biological science and medicine. Metrics for success are many, often highly subjective, and sometimes difficult to validate without historical perspective. However, administrative structures and policies that generate a scientific culture characterized by mutual respect, support, and enthusiasm among faculty and towards the institution as a whole, must go a long way toward achieving such success.

TWO FUNDAMENTALLY DIFFERENT PARADIGMS FOR RESEARCH DEVELOPMENT

The approach taken in this exercise is to define alternative organizing paradigms for developing research activity at biomedical research institutions, and then to contrast the implications of each of these choices across spectrum of elements that comprise the operating structure of the enterprise.

The philosophy and a selection of supporting policies used in developing the faculty in basic science at the Hutch, contrasted with a widely employed program-driven approach to both basic and applied research development in academic institutions, are used as examples that might usefully inform decision making at such institutions in general. An overall description of the alternative models may be given as follows:

Faculty-based development is driven by the perceived talent and productivity of individual independent faculty members within broad goals for the group and institution. The specific program of selected faculty is secondary. Areas of research concentration are targeted by faculty interest. Faculty members are grouped in broad scientific categories conducive to spontaneous intellectual interaction and collaboration. Governance is a shared responsibility, which includes selection and career development of faculty by peer review.

Program-driven development is focused on specific and more narrowly defined problems or fields lead by senior scientists. Programs are strongly vertically integrated, are specific goal oriented and frequently emphasize team rather than individual research. Selection and retention of research faculty is made by program heads to meet specific technical and intellectual needs and program goals.

SELECTION OF FACULTY INVESTIGATORS

Given the laboratory and ancillary space, and budgetary resources required to recruit and establish a new faculty member, the principal effort in faculty-based development is to identify scientists who, based on their accomplishments so far, are likely candidates to play leadership roles in their field of interest within the broad goals of the institution.

The recruitment procedure may begin with a list of preferred fields described in advertisements in leading journals. The breadth of the advertised list is an expression of the judgment of the faculty as a whole. The focus of the selection process is on the talent and perceived promise of applicants as leaders in their field, as well as the "fit" of the candidate within the makeup and culture of that faculty.

In the program-driven paradigm, selection of a specific field to target, along with decisions to allocate space and resources and initiate faculty-level recruitment, usually derive from the vision of senior leadership, often one person. While a search committee with independent faculty membership and national advertisement of the open position may be employed, the charge to the search committee comes from senior leadership.

Selection of successful candidates lies strongly in the hands of that leadership, principally on the basis of perceived needs of the program, even if arguably stronger candidates outside of the specific program focus are turned up by the search.

ROLE OF FACULTY VOTING.

Actions that are broadly supported by the judgment of the faculty form the core of peer review-based decision-making in faculty-based development. No action is more central to this process than voting on new membership. A large majority of the voting unit of the faculty (at least 75% in the case of the FHCRC Division of Basic Science) must support the proposed appointment in order to transmit a positive recommendation to the institutional executive (the President and Director at the Hutch). The final decision does lie with the senior executive.

If all of the agreed policies and procedures for faculty appointments have been properly followed, and required salary, space and resources are in place, then senior leadership will rarely exercise its authority to make a negative decision. Optimally, leadership works with the faculty to facilitate the proposed recruitment in what is usually a highly competitive marketplace for top talent.

Faculty and/or search committee voting may also take place with program driven development. It is understood, however, that program and senior leadership must be fully satisfied by the choices made. In some cases a short list of reasonably attractive candidates may be presented for final selection by leadership.

In the case that a search committee fails to identify a candidate acceptable to leadership, they may simply be thanked for their service, dismissed, and another search process initiated. Of course, even a faculty-driven search process may fail to identify a candidate who generates broad and enthusiastic support, with the same result.

COMPARATIVE ADVANTAGES AND LIMITATIONS

Meaningful comparison of these alternative approaches requires knowledge of both the starting point of development and the extent to which program-driven activity is combined with faculty status and career development. Program-driven development makes sense when an institution is first started and initial recruitment is built around established scientists who must pioneer the development of the institution.

For example, initially, the Hutch developed around a set of specific programs lead by a handful of senior scientists. About five years after the opening of its research facilities, a faculty-driven model was introduced as an option for further development of the institution. The critical distinction was that faculty appointments and career development were based on individual scientific achievement as judged by the faculty rather than primarily by specific program leadership.

This policy provided a broader base of professional scientific judgment across logical major divisions of the developing institution, such as the faculty of the Division of Basic Science. Importantly this shared responsibility for core decision-making provided the glue for knitting together a highly effective scientific culture. We found that faculty members take their votes seriously, and to a large extent, buy into a sense of responsibility for the success of new faculty appointments.

This sense of community extends well beyond the specific research interests of individual faculty. There is a continuing interest in what is going on in colleagues' laboratories, a willingness to interact and advise, and enthusiastic cooperation with the various mechanisms and events, described below, that are intended to promote cohesion within the community.

This approach is particularly effective for talented junior faculty members who have independent aspirations and may chafe at real or perceived limitations in a program structure. Truly groundbreaking research is often achieved by scientists relatively early in their careers. Hence, the faculty-driven model prefers recruitment of young scientists who show promise of pioneering such groundbreaking achievement as their career develops. It is precisely this type of faculty member who, when the hoped for success occurs, may not be attracted by or cannot be persuaded to remain within a program-driven structure. This is particularly true if a faculty appointment, and access to institutional space and resources depends upon adherence to a specific program.

Program driven development does have advantages early in institutional development, and when initiating major expansions of the research portfolio. Senior established investigators with extensive external grant support might be persuaded to play a leadership role by offers of ample space, ancillary resources and the opportunity to direct recruitment of additional faculty to fill out the program.

An integrated plan focused on a topic highly relevant to current trends in specific fields may readily be explained to development officers, boards of trustees and community donors. Hence institutional budgetary support may be generated for program needs for faculty salaries, capital expenditures, and other challenging financial goals.

In contrast, development focused primarily on faculty talent rather than program topic may present a steeper challenge for institutional development officers and a more complex case to make to lay board leadership and community donors. Generating support for the faculty-based model requires a more sophisticated, nuanced understanding of the frequently non-linear path of scientific progress. Furthermore, faculty team science requiring a program structure may be more or less essential to some types of research. An example is clinical exploration of novel problems and/or intensive therapeutic experiments utilizing specialized patient care facilities.

The marrow transplantation team at FHCRC is a prime example of sustained, highly successful, program-driven research development. Obviously every faculty investigator could not have his or her own intensive care marrow transplantation facility. The marrow transplantation program provides an impressive demonstration of the advantages of program driven development at a biomedical research institute; however, it is a unique story that is not easily duplicated.

Considered more broadly, program-driven development may appear to be very attractive at its outset. Progress in science, however, can move rapidly. What seems cutting edge at one point often becomes obsolete over surprisingly short periods of time. What happens if importance fades for types of faculty interests that were initially included in a program? A risk of faculty recruitment primarily for program-based needs is that individuals dependent on program achievement may prove unable to maintain productivity as, inevitably, science moves on.

This hazard for sustained scientific excellence at an institute increases with long term institutional commitments made to programs for faculty positions, space and resources. In contrast, in the faculty based model, investigators recruited for their independent accomplishments are likely to be adaptable to progress and change in their field, and to continue their individual productivity as a manifestation of the scientific talent that got them their job in the first place.

Admittedly, this contrast between developmental models can be too simplistic. Faculty recruitment to specific areas of research (i.e. programs) can meet high standards for individual talent and capacity for independent research. Furthermore, independent faculty can, and ideally should, form shared-interest groups where cooperation and collaboration is advantageous for optimal progress. Such groups can take advantage of funding opportunities for program grants and the development of resources beyond the scope of individual labs.

Highly productive groups can evolve into "empires" with the advantages and risks described for initially planned program development, especially if the group is successful in obtaining extensive funding resources. A policy of discouraging new faculty recruitment from the internal post-doctoral pool may serve to limit "empire" building. The policy of limiting laboratory size of individual faculty members (described below) may have the same modulating effect. When the advantages of such groups become less compelling over time, in the faculty based development model, they can be dissolved more easily.

Development at the Hutch included large elements of patient-based clinical, population-based public health, and laboratory-based fundamental and translational research. As mentioned, after an initial program driven period, basic science developed and continues under a robust faculty-based model. Research in the other elements developed and continues with a mixture of faculty-based and program driven administrative structures.

Retention and Promotion of Faculty

One of the advantages of faculty-based development is that it provides a well-understood and orderly basis for career development. Periods of appointment, institutional commitments and expectations of performance for junior and senior faculty are made explicit. Mentoring by senior faculty is made available and encouraged for junior faculty in order to provide advice and counsel with regard to both internal (retention, promotion) and external (grantsmanship) performance reviews. Decisions concerning promotion and retention of appointment are based on peer review and faculty voting. A general principle at the Hutch is to be as rigorous as possible in making initial appointments, and then to provide as supportive an environment as possible to promote the success of faculty members.

A prime example of success would be for the productivity of entry-level junior faculty to qualify for promotion to senior rank (generally requiring sustained national and international recognition and leadership in their field of research). The general philosophy at the Hutch has been to pursue career-long advancement for each faculty member as opposed to a "weeding out" competitive process.

A benefit of this approach has been to promote a climate of mutual support, and to reduce unproductive competitive conflict among the faculty. In addition to adherence to rigorous standards, an admitted challenge of this approach is its requirement for careful planning with respect to space and resources in order to provide for healthy growth of junior faculty labs, faculty turnover and the periodic introduction of "new blood".

Central to this discussion are methods employed to evaluate faculty performance. Objective metrics for this purpose are in widespread use, for example quantity of grant support, numbers of peer-reviewed publications, and "impact factors" such as citation frequencies. The value of some of these techniques for comparisons between whole institutions or scientific journals can be debated. Their application to individual faculty scientists, while providing some information, is a decidedly incomplete approach to reaching fully informed decisions.

An example of such a deficiency is the unfortunately common use of the Journal Impact Factors of a scientist's publications as surrogate for the quality of a candidate's published research. The San Francisco Declaration of Research Assessment (DORA) is a recent expression of the shortcomings of this approach that calls for the elimination of this practice (http://am.ascb.org/dora/). Many leading scientists, research journals, and institutions have endorsed it.

The fully developed picture of a candidate's progress, status in his or her field, and prospects for continued success are developed by external evaluations of a candidate's accomplishments by a fairly large panel (10 to 15) of reviewers made up of leaders in the relevant field(s). This survey is followed by a thorough discussion of this record by eligible voting faculty. A faculty vote (by faculty above rank in the case of promotions of junior faculty) is the definitive method of communicating a result. In our, and most other, similar institutions the form is of a recommendation to the institutional executive for final decision and action.

Once a faculty member achieves senior rank it still remains important to review and document continued productivity. On one hand, respect for the sustained achievement required for promotion to senior rank is important for morale and a reputation for fairness. On the other hand, lifetime sinecures for senior faculty are also not appropriate for the health of a research institute.

Full service universities may have opportunities for valuable activities, such as administration, teaching and community service, for senior faculty whose research productivity has permanently diminished. There may be, however, relatively little for faculty to do besides research at a dedicated research institute.

At the Hutch, peer-review of research by senior faculty members in basic science is conducted at five-year intervals by senior leadership with the help of an external peer panel. Results and recommendations are discussed with the faculty member. Plans going forward are tailored to individual circumstances. There are firm limits. Persistent failure to raise grant resources sufficient for a vigorous and competitive research program leads to loss of position for all faculty members.

This approach to faculty career development, in rigorous application, fits with faculty-based development. In the program-driven model some elements and forms, such as faculty and/or committee voting and external peer review are also frequently employed. The role of program leadership is robust and usually more determinative in final decision-making.

ROLE OF LEADERSHIP

Leadership in the program-driven model is fairly straightforward. Program leaders are essentially chief scientists directly responsible to the institution for the overall scientific success of their program. The faculty-based model also requires effective leadership, but of a more distributed and nuanced nature. Decisions need to have broad-based support within the faculty promoted by wide consultation and demonstrated by voting where appropriate.

At the Hutch, this post is called a Division Director (in either paradigm). The Division Director manages the process of faculty recruitment and career development, generates Divisional budget proposals, and acts as a spokesperson and administrative bridge between the Division faculty, the institutional administration, and other faculty units of the scientific community.

Division Directors in all cases are appointed by, and serve at, the pleasure of the President and Director of the institution. To be effective in the faculty-based model, and usually in a program structure, they also require the respect and trust of their faculty.

Distribution of Resources

Space and Size

There are two size related issues to discuss, particularly in relation to a faculty-based organization. These are the size of faculty voting units (e.g. Divisions or Departments) and the size of individual faculty laboratories within these units. With respect to the former, a goal is to develop and maintain a faculty size and physical proximity small enough to promote both knowledge about the research of colleagues and strong professional relationships between faculty members.

This need for professional interaction needs to be balanced by an overall faculty size large enough reasonably to cover the scientific interests and presence of needed expertise for the group as a whole. I'm not aware of any quantitative research on this issue. Our experience in Basic Science at the Hutch suggests, at least to me, that a target Division strength of up to 30 faculty members provides for both a cohesive faculty culture and sufficient "natural" faculty turnover to sustain needed change as science progresses over time.

The size of individual faculty laboratory groups varies widely among and within biomedical research institutes. These groups range from large (e.g. "25 post-doc") laboratories occupying whole floors to small labs with just a few post docs, students or technicians sharing just two or three laboratory modules and an office. The large laboratory is common in the program-driven model of development, particularly for the senior leadership of the program.

In contrast, in the development of Basic Sciences at the Hutch, we employed a specific formula for assignment of space for individual faculty and their groups. Entry-level Assistant Members (equivalent to Assistant Professor) were assigned a three-module (or three bay) laboratory of about 750 sq. feet. With promotion to Associate Member an additional lab module was added to accommodate growth of the program. An additional fifth module upon promotion to full Senior Member status followed this promotion-based space assignment policy.

Even with five modules this constraint in laboratory space tended to keep group size limited, the Principal Investigator close to the bench and the experimental work, and to his or her students and post docs. The faculty member is often the most effective experimentalist, but supervising very large groups may diminish opportunity for creativity and tend to drown the Principal Investigator in administrative detail.

OTHER RESOURCES

Beyond bench and office space needed for setting up and maintaining faculty laboratories, there are other resources; including laboratory modification, specialized facilities, and expensive items of equipment generally out of reach of individual grant budgets. Our approach was to deal with such needs and requests for institutional support, so far as practical, as an automatic part of faculty evaluation procedures accompanying recruitment and promotion.

In our faculty-based model of development this policy provided the institution with expert peer-review for distribution of space and resources and assured every faculty member that their needs would get serious consideration within the Division without the need for (and in fact discouraged) special pleading or private lobbying of divisional or institutional leadership.

That said, it remains important to recognize outstanding performance. We found that such performance can be rewarded by the timing of performance reviews and resulting recommendations for increases in space, resources and salary.

In my opinion, the sense of fairness and trust fostered by these policies contributed significantly to the ability of faculty to focus more energy on their research and less on internal politics. Academic biomedical research is a tough, highly competitive enterprise, and has been only getting more so in recent years. A useful goal is to develop a research institution with an internal environment that is seen by its scientists, in so far as possible, as part of the solution rather than part of the problem.

SHARED CAPITAL RESOURCES

A community of scientists, in an up-to-date biomedical research institution, need access to large-scale resources and equipment. These vary from research libraries and animal care facilities, traditionally provided to all members by the institution, to an ever-growing and evolving list of capital-intensive resources and equipment based in advancing technology.

Tools for research in genomics and proteomics, mass spectrometry, biological imaging, monoclonal antibody production, flow cytometry, microarray and related screening technologies, bioinformatics and platforms for data analysis, histopathology, and specimen processing and storage, among others, are current examples of such shared resources in a modern biomedical research institute.

Typically such resources are available generally to the faculty and their laboratories, have PhD-level managers, technical staff for maintenance and assistance to users, and are supported by user fees charged to the research grants of faculty users.

Support for this type of shared resource is a major element of Cancer Center Support Grants (popularly called “Core” grants) from the National Cancer Institutes (NCI) and a significant benefit to investigators in participating cancer research centers. From its inception in the mid-1970s, the NCI Core grant system has played a significant role in developing this approach, which is now widespread in research universities and institutes of all kinds in the US and internationally.

The shared resource approach fits seamlessly into the faculty-based model of development. Programs (and very large individual laboratories) can also utilize and benefit from shared resources. When such resources are intrinsic to and/or developed within a program or large laboratory, however, access to that technology may or may not be available to outside investigators.

INTERIM AND BRIDGING FUNDING

The peer-reviewed grant system of the National Institutes of Health (NIH), and other similarly governed granting agencies, is central to the overall success of the national biomedical research enterprise. Faculty success in obtaining competitive funding from this source is also central to maintaining quality control in both faculty-based and program driven models of research development.

Exclusive reliance on this system, however, may fail to promote the highest levels of scientific achievement, especially in periods of serious and sustained budgetary constraint, such as we are currently experiencing.

As national competition between scientists for limited research dollars increases, decision making by grant review committees necessarily grows more conservative, and to some extent more arbitrary. Novel ground breaking research proposals, necessarily attached to greater risk and uncertainly, may be particularly vulnerable to being passed over by the orthodox national peer review process.

Therefore, it falls to research institutions that aspire to a leadership role to develop and maintain financial resources in support of such important efforts. Institutional interim and bridge-funding policies, and sometimes pilot funding programs, can provide vital support until sufficient progress, and the passage of time, eventually lead to success in conventional external grant funding.

The reputation of locally obtained research funding, outside of the peer review system, has at times, and quite rightly, been suspect as a pathway to clogging institutions with pedestrian, largely unproductive, research activity. Therefore rigor in decision-making with regard to this type of funding is vital and needs to be thought through carefully. In times of financial stress there may be no other more important need for a research institution to address.

ENHANCING SCIENTIFIC CULTURE

Virtually every biomedical research institution worthy of the title has programs and policies intended to foster communication, cooperation and collaboration among the members of its scientific community. Popular examples are scientific retreats, internal and external seminar series, mentoring mechanisms for graduate students, postdoctoral fellows and junior faculty, scientific interest group, club and/or literature review meetings, and the local organization and sponsorship of national and international scientific meetings in relevant fields. These activities help enrich scientific culture and maintain the intensity of scientific professional life that characterizes top research institutions.

One such exercise, a weekly lunchtime faculty seminar series called "Faculty Lunch", played an important and sustained role in faculty-based development in basic science at the Hutch. I have not seen this program employed very often elsewhere.

The content of this weekly lunch hour meeting consists of a presentation, by a faculty member to the Divisional faculty, of one or more facets of current research in his or her laboratory. The schedule of assigned presentations is set at the beginning of the academic year. Trading dates in order to accommodate busy faculty schedules is fine, but attendance is understood to be a faculty obligation. The schedule runs until every faculty member has presented, usually by the end of the academic year.

While most, if not all, of the Hutch's internal and external meetings and seminars are open to all members of the local scientific community, "Basic Science Faculty Lunch" is focused specifically on, and for, faculty members as a core mechanism for scientific and cultural cohesion. In faculty-based development it is difficult to command participation in any type of seminar program.

Enthusiasm for participation tends to wax and wane over time. In contrast this exercise has been maintained, through changes in leadership and over the several decades' long history of the Division, despite the many other demands on faculty time. Faculty Lunch was, and still is, seen as a major value to the scientific life of the Basic Science faculty.

A reason for this cultural success may be the surprising degree to which modern biomedical scientists from distinct fields of research, use similar concepts and tools, and face similar technical and intellectual challenges. The Basic Science faculty at the Hutch, by way of example, comes from many different traditional disciplines, attends different national scientific meetings, and often publishes in different specialty journals. Individual members have been chosen in part to bring an added dimension to the scientific program of the Division, rather than an overlapping and potentially competitive environment.

Faculty Lunch provides overviews of current work and progress from colleagues who command respect in distinct fields without the competitive edge that sometimes dominates meetings in their own fields. I and, I believe, many of my colleagues got fresh ideas and perspectives for their own work from this exercise. Moreover, the lively discussion characteristic of Faculty Lunch provides help and advice to the presenting faculty member from an audience different from one that he or she usually addresses.

Formal scientific collaborations and co-publishing are often taken as evidence of strong cohesion within a faculty, and thought to promote effective leveraging of talent and expertise. Program-driven development is often built around formal collaborations, such as program-project grants, which are the centerpiece of the research enterprise.

Formal collaboration can and certainly does occur in the faculty-based model of development. In that case the cherished independence of faculty investigators means that a compelling scientific (and sometimes economic) rationale needs to be present in order to stimulate and maintain voluntary formal collaborative research.

Often overlooked in external reviews and critiques of faculty collaboration are the myriad ways in which cooperation, peer education, sharing of knowledge and technology and so forth, within a well-functioning faculty environment, can provide benefits approaching those of formal collaboration without requiring co-publication. Formal collaborations and co-publishing may arise in such settings as a result of "spontaneous combustion" among faculty laboratories rather than from top-down direction in a program structure.

Furthermore, scientific progress attributable to individual scientists is as significant as that attributable to groups. It must be admitted, however, that given the fluid and nuanced nature of collaboration within an independent faculty, demonstration to external institutional reviewers of an effectively collaborative faculty-based enterprise sometimes presents a challenge.

PREDOCTORAL AND POSTDOCTORAL TRAINING

Faculty sometimes prefer careers in research institutes in part because they may be able to minimize a teaching burden seen as a distraction from research. Objectively, however, doing and teaching science are deeply intertwined. For example, post-docs and graduate students provide the vast majority of the manpower for the academic research enterprise. A robust post-doctoral research program that brings talented trainees together with effective faculty mentors serves as a strong essential component of an optimally functioning biomedical research institute.

In contrast, PhD-level graduate training can be a subject of contention, especially between full service universities and affiliated research institutes lacking independent degree granting authority. The leadership and/or faculty of an academic university may view their privileges, duties, and obligations quite differently than do their counterparts at a research institute and, therefore, find it challenging to share a graduate training program. At the Hutch, it took a decade and a half of, at times, frustrating discussion, and several false starts, to come successfully to an agreement on a joint interdisciplinary graduate program in cellular and molecular biology (MCB) with the University of Washington. The result, well worth the effort, has been every bit as rewarding as was envisioned by optimistic advocates for the program at both institutions.

Post-docs are young scientists developing their long-range interests and attempting to establish a track record of accomplishment sufficient to enter the job market and earn them a faculty position and their own lab. In contrast graduate students are learning to be scientists by enlarging their scientific knowledge, technical mastery, and establishing their ability to design and execute experiments. The kinds of instructional needs and questions presented by students to their faculty (and often post-doctoral) mentors require regular review, and sometimes rethinking, that serves to challenge and refresh. A balance of students and post-docs is a valuable asset for creative productivity in a faculty-lead group in a research institute.

Furthermore the shared responsibility of managing a graduate program promotes faculty cohesion and can contribute to high scientific standards for the institution. Unlike post-docs, who are recruited and hence quality-controlled by individual faculty members, standards for talent and continued performance of graduate students can be set and maintained by the faculty as a whole.

The competitive success or failure of annual recruiting for top students can provide valuable information on the scientific standing of an institution. Finally, the achievement of the cross-institutional MCB platform for a joint program in graduate training has served as a template on which to build similar joint programs, within MCB, for other University of Washington affiliated research institutions in the local scientific community.

Relations between Faculty Investigators and Institutional Administration, the Board of Trustees and the Wider Community

Outside of working scientists, it takes many people to develop and run an effective research institution. A substantial reference work on research administration was mentioned at the outset of this discussion[1], but even this tome doesn't cover all the important supporting elements that make a great research institution. An incomplete list of examples not covered includes innovative applications of library science, advanced systems of information technology, and programs of educational outreach to the general community, such as a Science Education Partnership that connects Hutch scientists with local educators.

I do not in any way mean, by skirting these important topics here, to minimize the vital contributions of employees, administrators, development officers and staff, and supporters from the community. For there to be sustained success a deep sense of mutual respect, gratitude, and partnership must exist between the professional scientific staff, and all of the other participants in the enterprise. Attention to mechanisms of communication between all participants is a core function of institutional leadership.

SUMMING UP, THE ROLE OF EXTERNAL ADVICE AND ADAPTATION TO CHANGE IN MAJOR TRENDS IN SCIENCE

So in the end, how does one summarize the comparative merits of the two organizational paradigms that I introduced at the outset of this exercise? It has been argued that, if transformational scientific progress is the goal, major innovations are more likely to arise from institutions that tolerate novelty and non-conformity to current thinking[3]. This notion may support the use of the faculty-driven model in research development in preference to the intrinsic rigidities of a program-driven structure, particularly over time.

There is, however, little or no formal scholarship that classifies and directly compares faculty-based and program-driven research development in generating major discoveries or any other milestones of scientific progress. Such research might be of value to institutions engaged in developing their research portfolio.

Furthermore, the time scale for institutional decision-making about key issues, such as whom to hire, and on what fields of research to focus, may not be compatible with the time it takes to recognize historically important innovation. As a practical matter, development requires real time inputs in order to reach the most informed decisions possible. The quality of those decisions, in terms of a major impact on scientific progress, can usually only be assessed in retrospect.

So other inputs are required to inform evaluation. I have already addressed peer-derived information used to evaluate individual candidates for recruitment and promotion. For overall institutional performance, and in addition to conventional metrics of quality (e.g. numbers and "impact factors" of publications, success in peer reviewed grant support, awards and other recognitions to faculty members) there are sources of peer-derived information that provide a real time overall assessment of how well a research institution is doing.

Experiences with competitive recruiting of top-level candidates for faculty positions, and even graduate student recruitment statistics, can be eye opening in this regard. The Hutch has found that external review boards composed of scientists, held by both faculty and leadership in high esteem, can be of substantial benefit.

To be effective, however, such exercises must be carefully organized, directed to issues of real institutional significance, and respectful of the valuable time of both the reviewers and the reviewed. What are not helpful are imposed review exercises, held primarily as window dressing, in which neither the institutional leadership nor faculty has any serious intention of responding to recommendations. In addition to the formal written product of such reviews, usually couched in (and blunted by) carefully worded diplomatic language, I have found that opportunity for informal conversation with reviewers helps get the message across and enhances the useful impact of the review.

Among the most pervasive trends in biomedical research today are efforts to accelerate translation of discoveries in basic science into effective new treatments for human diseases. Much of this activity reflects attempts to harvest the "low hanging fruit" of more untargeted discovery-driven fundamental research in biology. The goal is certainly laudable, and some success has been achieved such as in new, and more personalized, clinical applications in cancer treatment.

The current emphasis on translation, however, has its drawbacks. A recent editorial in Science[5] points out that overall progress is slower than hoped, most likely because fundamental knowledge is still lacking. The opinion expressed is that new and continuing basic research is needed to generate opportunity.

The focus on funding translation seems to be the current iteration of the more general problem of productively managing the relationship between basic and applied research. A prescription for a national science policy was initiated by

Vannevar Bush in establishing the principles for government support of civilian research after World War II[6], and has been elaborated and critiqued in the 1990s by James Stokes in an influential book, called *Pasteur's Quadrant*[7].

Bush posited that the impulse for applied research would inevitably push out basic discovery unless policies were in place to protect and sustain basic research (which then, in turn, generated opportunities for application). Although Stokes argued, persuasively, that the relationship between basic science and applied research is more complex, interrelated and dynamic than a straightforward tendency to mutual exclusion, present circumstances do raise a warning. The sustained and growing constriction of federal grant support for biomedical science, is driving scientific talent away from basic research, thereby distorting a wise and needed balance between untargeted discovery and translational application.

How the issue of translation relates to the topic of this perspective may be perceived from the fact that applied and translational biomedical research frequently proceeds from straightforward assumptions that current concepts and technologies can be used in a linear fashion to achieve specific goals. Indeed, sometimes, as in the case of the marrow transplant program at the Hutch, they can. Together with the requirement, in many cases, for considerable manpower and/or large laboratories and other facilities, large budgets and a strong team approach, the program driven model of development tends to dominate in these fields.

A problem arises, and not infrequently, when current knowledge is seriously incomplete, and a linear progression of research efforts leads nowhere. Historically the pathway to many of the major scientific and technical achievement in medicine (and many other fields) is far from linear. The acclaimed books and TV series called "Connections," by the historian of science and technology James Burke[8], documents the often surprising and clearly non-linear chain of connections leading from discoveries in antiquity to many of the celebrated achievements of modern society.

Certainly anticipating such surprises is well beyond the scope and plan of most program-driven biomedical research. My point is not to oppose program driven research but simply to point out that there is plenty of reason to incorporate the flexibility of faculty based development, even in translational and applied biomedical components of a biomedical research institute.

I have had the great good fortune to spend decades of my career in close quarters with phenomenally successful examples of both program driven and faculty-based research development. I hold my colleagues in these enterprises in the deepest respect and admiration for their accomplishments. If I have a concluding message based on this long experience, it is to avoid policies and decisions that diminish opportunities for transformative science for the long run.

I would simply warn against an unbalanced trend away from basic research, independent faculty, and the faculty-based model of research development. This concern is especially acute in this period of constrained funding and enhanced competition for research dollars. I can testify that such development does promote, both directly and indirectly, every facet of scientific progress. Moreover it can provide for scientists, as it has for me, a richly rewarding professional career, and one that continues to attract the most gifted of our young people to this field so vital for our progress as a society.

References

[1] Dingerson, MR. University Research Development and the Role of the Chief Research Officer. In: Kulakowski, EC, Chronister, LU, eds. Research Administration and Management. Sudbury, MA: Jones & Bartlett, 2008: 63-72.

[2] Tornatzky, LG. Waugman, PG. The Academic Entrepreneur: Enabling and Nurturing a Scarce Resource. In: Kulakowski, EC, Chronister, LU, eds. Research Administration and Management. Sudbury, MA: Jones & Bartlett, 2008: 233-46.

[3] Hollingsworth, JR. Major Discoveries, Creativity and the Dynamics of Science. Vienna, edition echoraum, 2011

[4] Lotka, AJ. The Frequency of Scientific Productivity. J. Washington Acad. Sci. 1923; 16:317-23.

[5] Zoghbi, H, The Basics of Translation, Science 2013; 339:250.

[6] Bush, V. Science – The Endless Frontier: A Report to the President on a Program for Postwar Scientific Research. Washington, US Government printing office, 1945, reprinted National Science Foundation 1990.

[7] Stokes, DE. Pasteur's Quadrant. Basic Science and Technological Innovation. Washington, Brookings Institution Press, 1997.

[8] Burke, J. Connections. New York, Simon and Schuster Paperback Books, 2007.

Made in the USA
Monee, IL
04 November 2026

67387912R00020